Dinosaurs walked the earth a long, lo

They ruled the earth for about 180 million years before disappearing 65 million years ago.

This dinosaur footprint found in the LaSal Mountains of Utah is nearly 2 feet long.

But everything we know about dinosaurs has been discovered within the last 200 years by scientists called *paleontologists* (PAY-lee-un-TAHL-uh-jists), who find, dig up, and study *fossils*. Fossils are the remains or traces of animals and plants that lived in the distant past.

How are fossils formed? How did dinosaur bones survive for millions of years? Actually, not that many dinosaurs became fossils. That's because very special conditions must exist for fossils to form.

Some of the dinosaurs that died were buried quickly, perhaps by the mud of rivers, lakes, or swamps. Other dinosaurs died on land and were pushed into water by heavy rainfall. Bit by bit, bacteria caused their flesh to decay, leaving only the skeleton.

Often minerals in the water entered the bones of the skeleton and changed them to stone. Sometimes the mud or sand hardened into solid rock around the bones.

As millions of years passed, the land changed. Hills and mountains pushed their way up. Rivers changed course or dried up altogether.

This turtle-shell fossil was found by paleontologists in the dry soil shown in the photograph above. The land was once the bed of a river.

Many early dinosaur hunters found fossils lying on the ground. Now most paleontologists have to dig for fossils. Sometimes they use a bulldozer to scrape away the top layers of rock.

When they get closer to the bones, they may break up the rock with jackhammers.

Then they chip away at the rock with picks.

Before paleontologists completely free the fossilized bones, there is much work to do. They must first note the exact spot where the bones were found.

Photographs of the bones in the ground will show how they looked when they were discovered.

The ancient bones are often cracked and crumbly. The scientists give them a coating of chemicals (below) to keep them from falling apart.

After a fossil is out of the ground, it is carefully covered with layers of cloth soaked in wet plaster (below). As the plaster dries, it hardens and protects the bones as a cast protects a broken arm.

A bulldozer then hoists up the fossils and carefully sets them down on the back of a truck.

The truck leaves the site and heads for a laboratory.

The paleontology lab is usually part of a museum or a large university. Here workers carefully unpack and store the fossils.

Lab scientists finish the job of cleaning the fossils with fine picks or needles.

Sometimes they use chemicals to dissolve the rock that still remains around the bone.

Now that the bones are ready, dinosaur detectives work hard to fit them together to form a skeleton.

Once a skeleton is put together, finishing touches are added before it is displayed in a museum.

Knowing exactly what a skeleton looks like helps sculptors and engineers to build realistic, life-size models.

Paleontologists study fossils to learn all they can about the dinosaurs. What can they learn about dinosaurs from the size and shape of their bones?

Scientists studied this three-horn dinosaur skull carefully. The skull was nearly 7 feet long. One horn was short, but two of the horns were each 3 feet long. The scientists named the dinosaur Triceratops (try-SAIR-uh-tops), which means "face with three horns." At the front of its skull, Triceratops had a beak that it used to slice through tough plants.

The powerful back legs and spread-out toes of Triceratops are like those of today's rhinoceroses. Experts say that Triceratops probably defended itself by charging with great force into its enemies.

Paleontologists found that Struthiomimus (STROOTH-ee-oh-MY-mus) had thin back limbs that looked like the legs of an ostrich. Its two shorter front limbs ended in three long, clawed fingers.

Could Struthiomimus run fast? Paleontologists looked at the length of its back limbs and the way its bones fitted together. They then concluded that this small dinosaur had long, strong leg muscles. Based on this evidence, paleontologists believe Struthiomimus could run as fast as 50 miles per hour.

What did Struthiomimus eat? Sharp claws on its front limbs show that this swift hunter could catch small animals. But it could also grasp tree branches. So paleontologists now believe that Struthiomimus may have eaten both plants and animals.

Imagine what comes to mind when scientists look at the bones of the dinosaur they call Tyrannosaurus (tie-RAN-uh-SAWR-us).

They see a beast about 40 feet long, nearly 20 feet tall, and weighing 6 tons!

The experts study the enormous, 5-foot skull and the 60 razor-sharp teeth. This dinosaur was no plant eater. Tyrannosaurus very likely used its sharp teeth to tear into the flesh of weaker dinosaurs and other creatures.

Tyrannosaurus had short but mighty arms that could probably lift objects weighing up to 400 pounds. Perhaps Tyrannosaurus used its powerful arms to capture its victim before opening its huge jaws and sinking its razor-sharp teeth into the victim's body.

Paleontologists are professional dinosaur detectives. But they depend on amateurs—people like us—to help them find fossils. Farmers, miners, and road builders find dinosaur fossils while digging in the earth. Some people find fossils while hiking and exploring. Should this happen to you, jot down the location and report your find to a university or a museum. Happy hunting!